03. 界面与网页设计

Design+
全球设计精粹 | 第2辑
《Design+》编写组 编

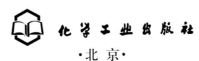

编写人员名单（排名不分先后）

许海峰	邓 群	张 淼	王红强	谢蒙蒙	董亚梅	任志军
张志红	周琪光	任俊秀	王乙明	胡继红	黄俊杰	柏 丽
袁 杰	李 涛	卢立杰	田广宇	童中友	张国柱	常红梅

图书在版编目(CIP)数据

界面与网页设计 / 《Design+》编写组编. —北京：化学工业出版社，2013.8
(Design+——全球设计精粹. 第2辑)
ISBN 978-7-122-18131-2

Ⅰ.①界… Ⅱ.①D… Ⅲ.①人机界面－程序设计－作品集－世界－现代②网页－设计－作品集－世界－现代 Ⅳ.①TP311.1②TP393.092

中国版本图书馆CIP数据核字(2013)第178393号

责任编辑：王 斌　林 俐　　　　　　　装帧设计：锐扬图书

出版发行：化学工业出版社(北京市东城区青年湖南街13号　邮政编码100011)
印　　装：北京瑞禾彩色印刷有限公司
880mm×1092mm　　1/16　　印张 11　　2014年 1 月北京第 1 版第 1 次印刷

购书咨询：010-64518888 (传真：010-64519686)　　售后服务：010-64518899
网　　址：http://www.cip.com.cn
凡购买本书，如有缺损质量问题，本社销售中心负责调换。

定　　价：68.00元　　　　　　　　　　　　　　　　　　　　　版权所有　违者必究

设计无国界

化学工业出版社建筑出版分社于2011年组织出版了《Design+——全球设计精粹》丛书,包含《平面设计》、《包装设计》、《企业形象设计》、《展览设计》《图案设计》、《网页艺术设计》六个分册,收录全球近100个设计公司5000余例近年间最优秀的平面设计案例,以极富创意的优秀案例和良好的成书品质受到读者的好评。一年多来,国际平面设计界又诞生了许多优秀作品,我们再次将这些案例集结成册,以飨读者。

延续前套图书的成功之处,本套丛书收录最新、最前沿、最优秀的国际平面设计案例,对其进行更为精细的筛选和更为合理的分类编排,分为《包装设计》、《画册 书籍设计》、《标志 名片与VI设计》、《广告 宣传页 卡片设计》《界面与网页设计》、《图案设计》六个分册。收录作品涉及全球约40多个国家及地区,旨在加强设计师间的国际交流,开阔设计视野,开拓设计思路,启发设计创意。除此之外,更注重版式设计和提升整体书籍装帧水平,改善阅读体验。

最后,再次感谢读者对本系列图书的支持,也希望这个系列的图书能越做越好。化学工业出版社建筑出版分社也将为设计行业专业人士、设计专业师生以及广大设计爱好者带去更多更好的专业图书。

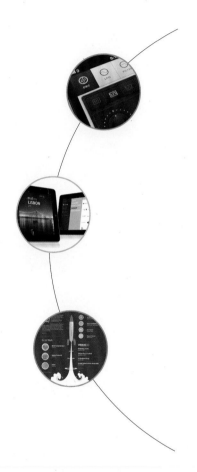

01. 设计元素 / 001

02. 平板与手机界面设计 / 009

03. 电脑网页设计 / 057

　　事业新闻 / 058

　　企业网页 / 070

　　电子商务 / 092

　　创意网页 / 120

　　综合网页 / 154

01.
设计元素
Design Elements

设计元素

512X512

256X256   128X128

界面与网页设计 INTERFACE & WEBPAGE

设计元素

005 +

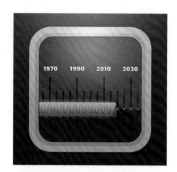

设计元素

平板与手机界面设计
Tablet And Mobile Phone Interface Design

界面与网页设计 INTERFACE & WEBPAGE

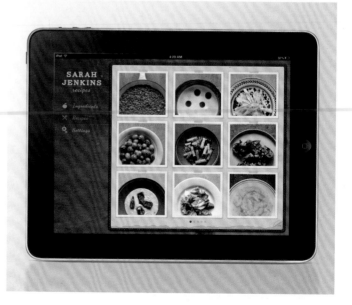

平板与手机界面设计

界面与网页设计 INTERFACE & WEBPAGE

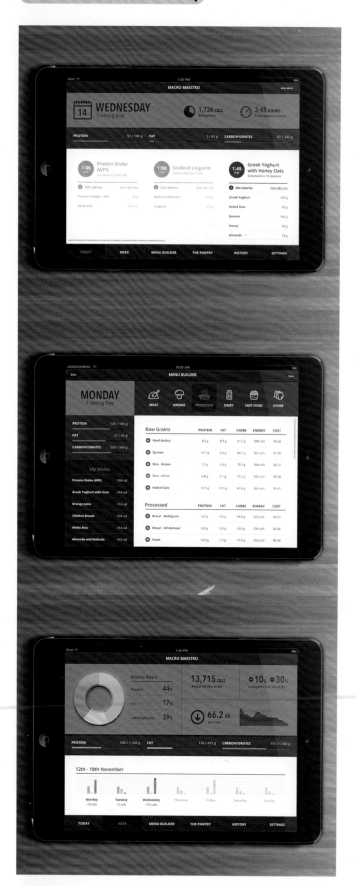

平板与手机界面设计

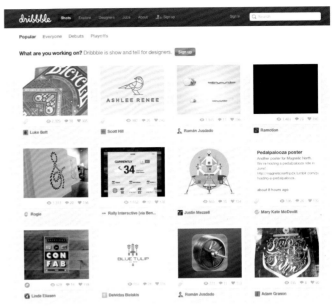

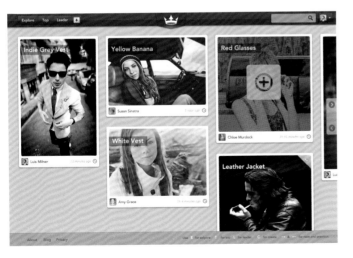
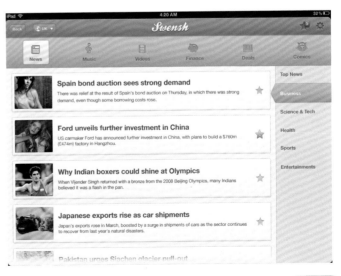

013 +

界面与网页设计 INTERFACE & WEBPAGE

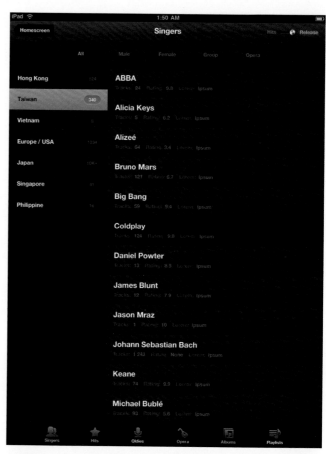
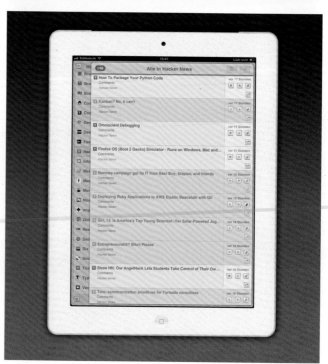

平板与手机界面设计

015 +

界面与网页设计 INTERFACE & WEBPAGE

平板与手机界面设计

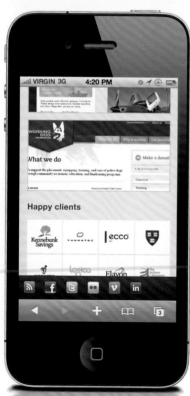

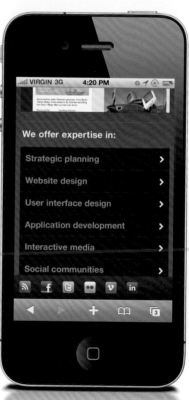

平板与手机界面设计

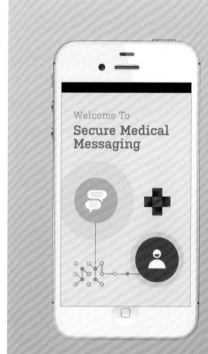
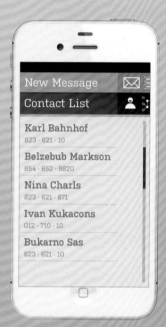

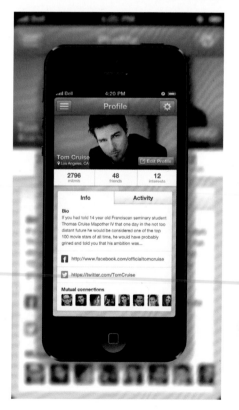

界面与网页设计 INTERFACE & WEBPAGE

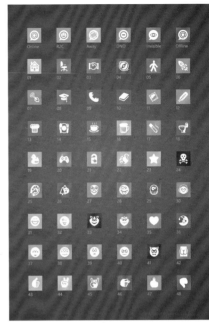

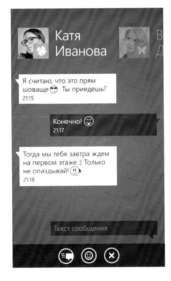
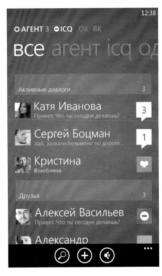
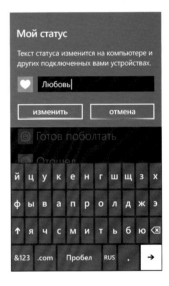

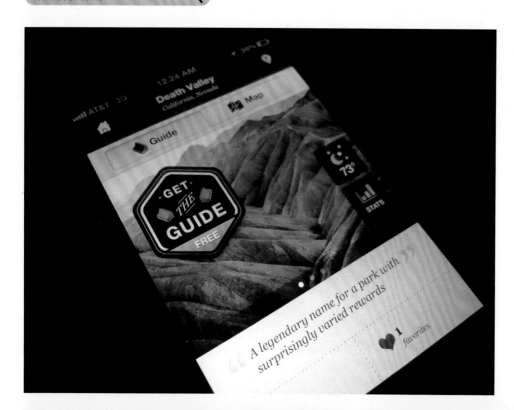

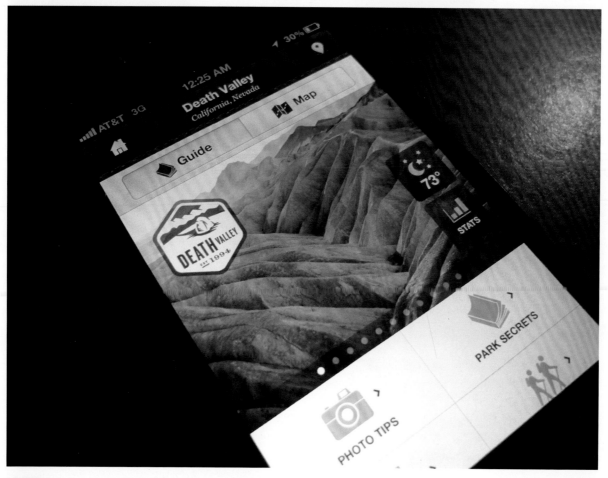

平板与手机界面设计

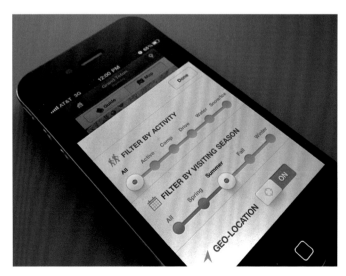

027 +

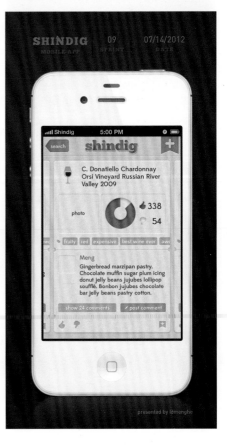

平板与手机界面设计

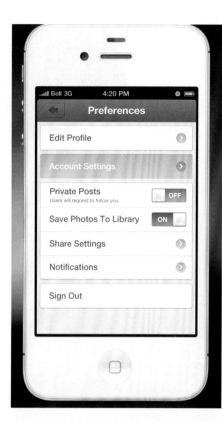
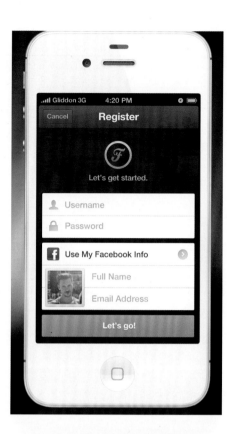

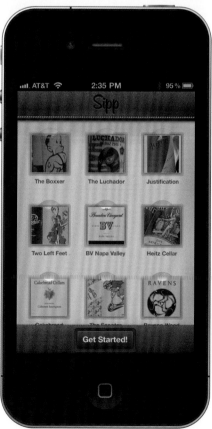

平板与手机界面设计

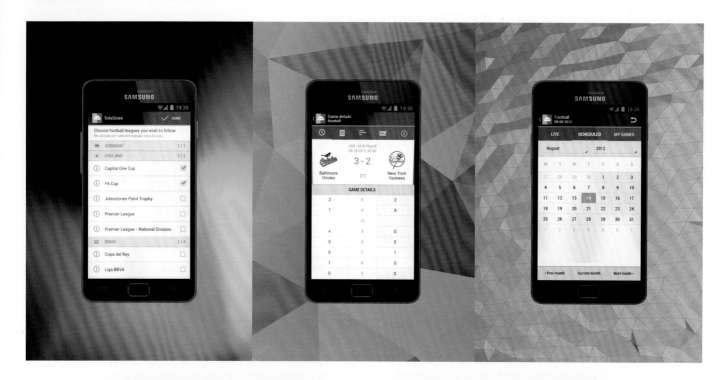

平板与手机界面设计

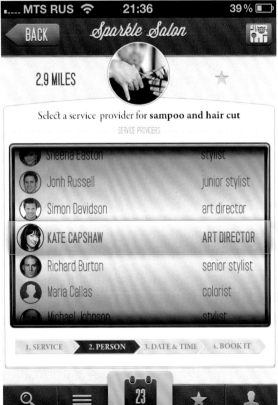

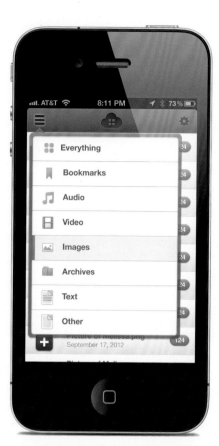

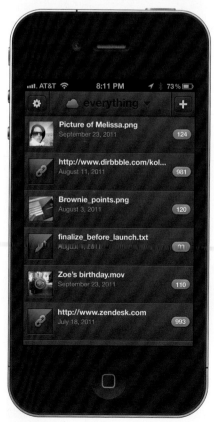

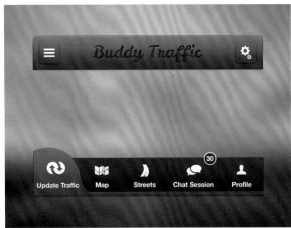

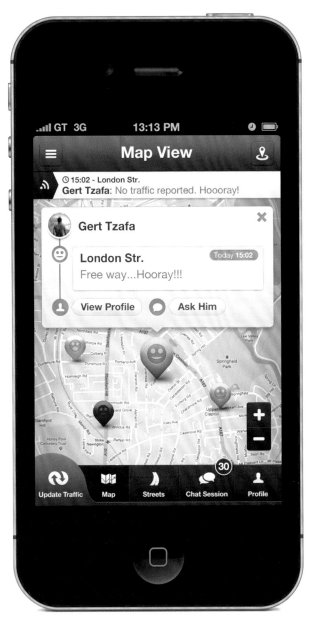

界面与网页设计 INTERFACE & WEBPAGE

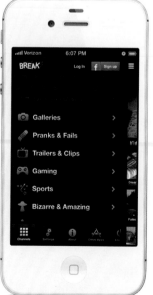

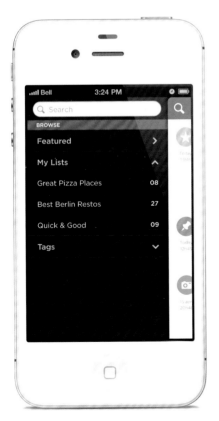
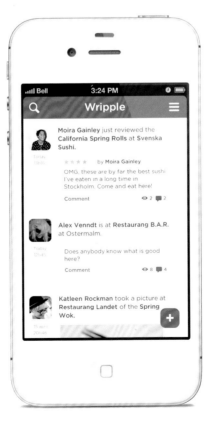
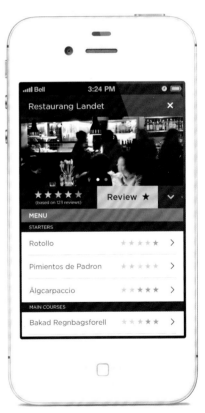

界面与网页设计 INTERFACE & WEBPAGE

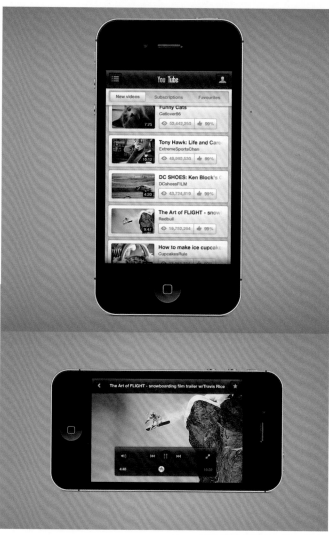
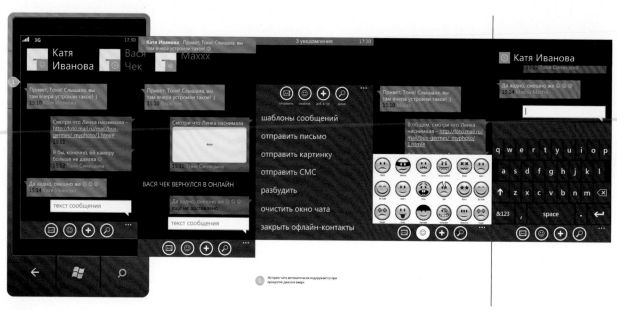

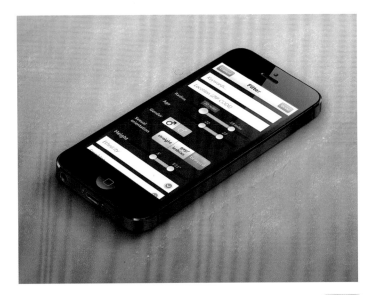

界面与网页设计 INTERFACE & WEBPAGE

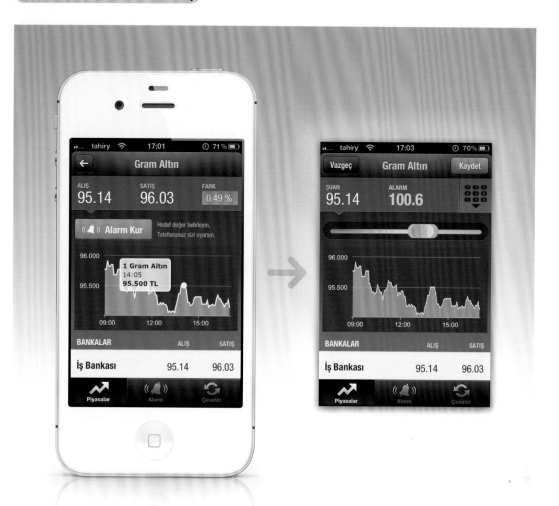

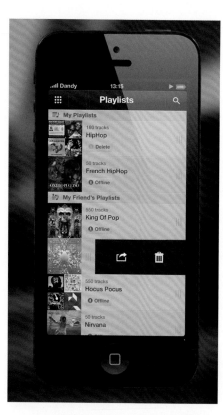

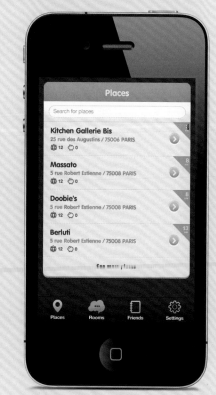
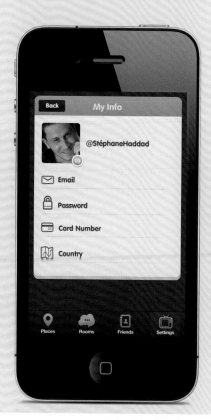

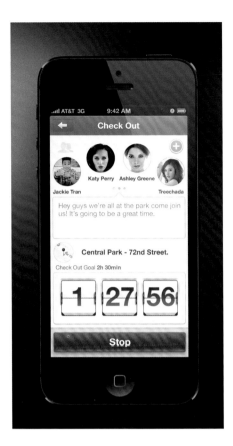

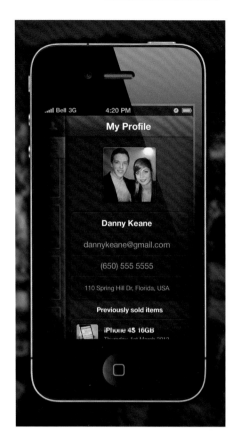
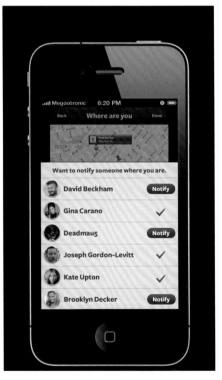

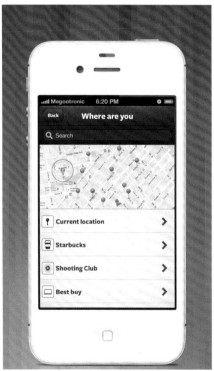

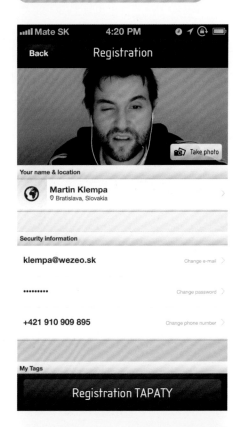

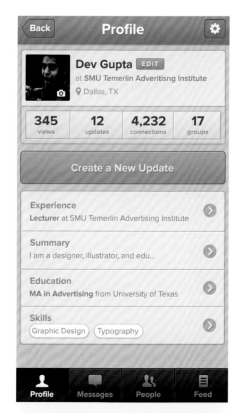
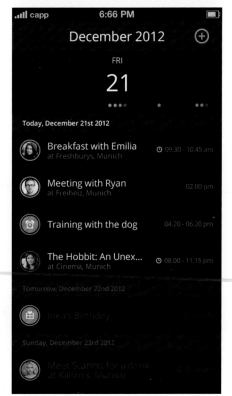

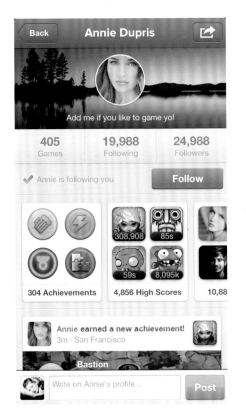
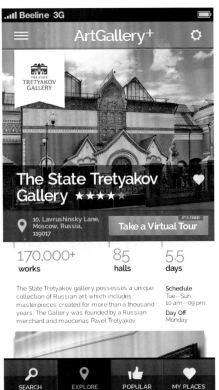
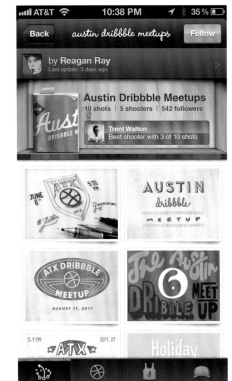

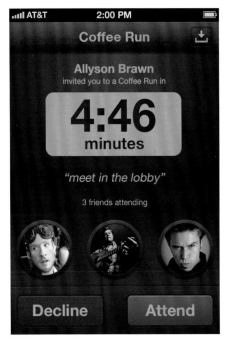

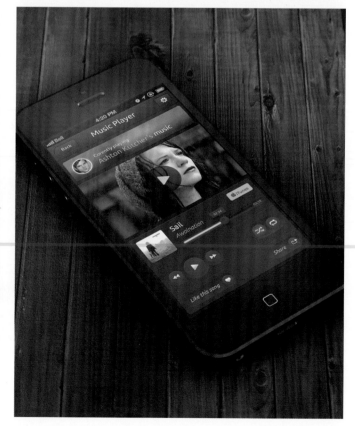

平板与手机界面设计

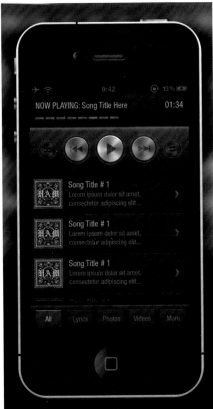

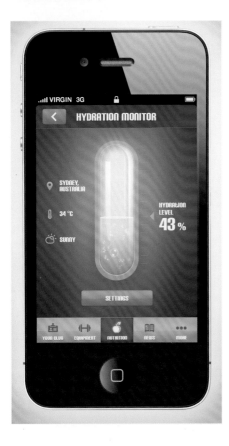

平板与手机界面设计

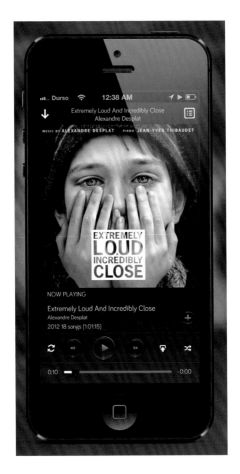
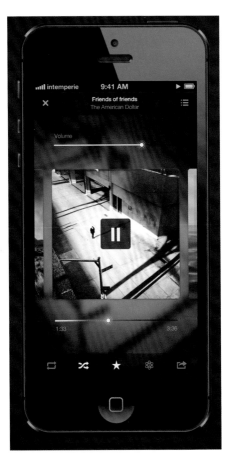

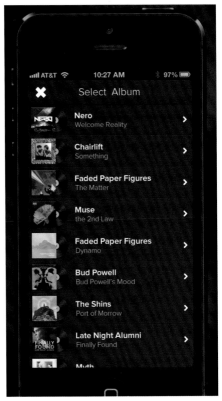
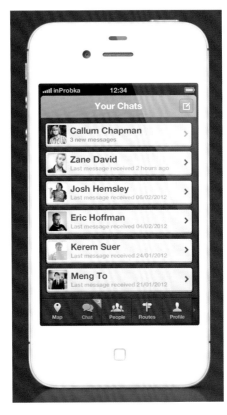

界面与网页设计 INTERFACE & WEBPAGE

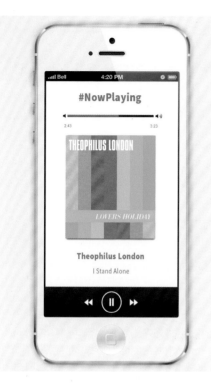

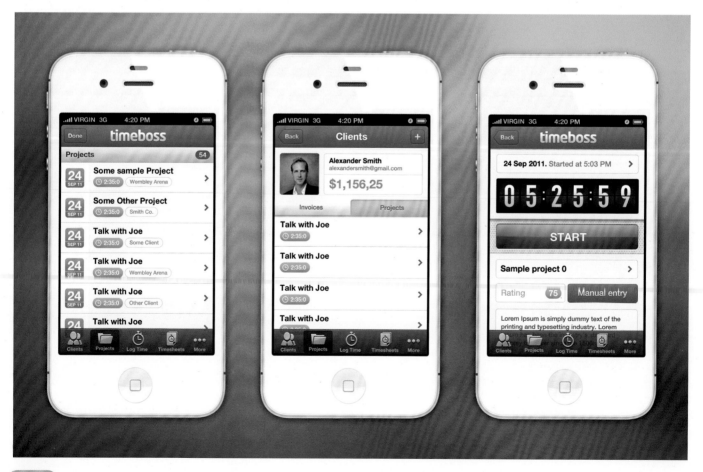

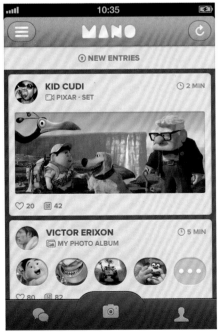
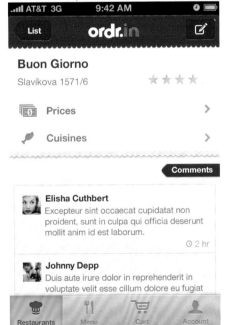

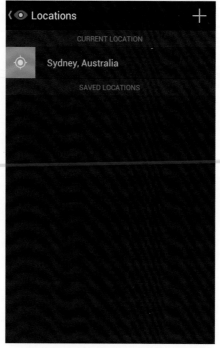

平板与手机界面设计

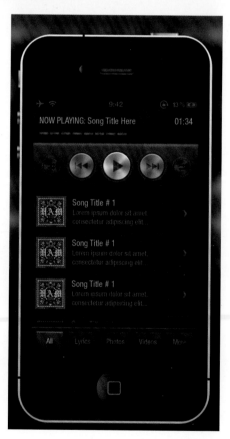

03.

电脑网页设计
The Webpage Design

电脑网页设计
事业新闻

http://industryconf.com/

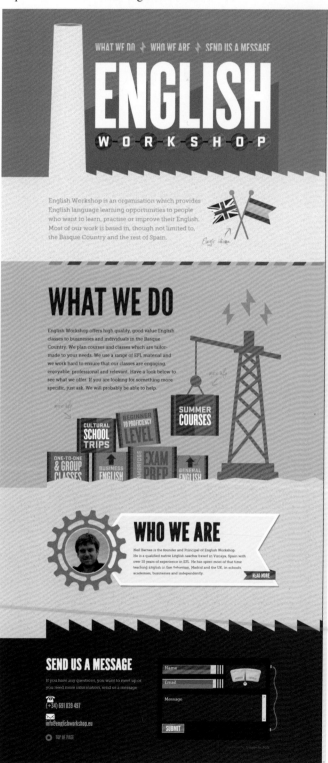

http://www.cursos-verano-inglaterra.com/

http://enblaze.co.uk/

http://fordoing.co.uk/

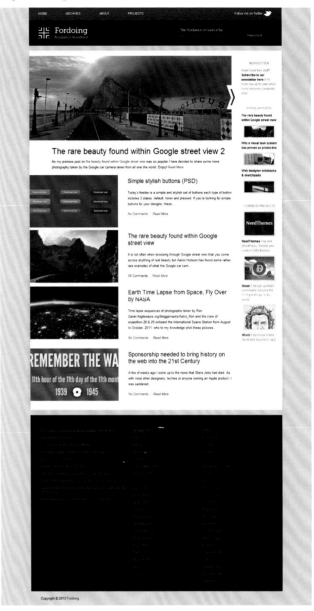

http://australia.gov.au/

http://joeycadle.com/

http://www.fukuokabank.co.jp/

http://www.sabey.com/

http://www.ryukoku.ac.jp/

http://www.kindai.ac.jp/

http://redfeather-photography.com/

http://phar-ma.com/

http://abcnews.go.com/

http://magazineworld.jp/

界面与网页设计 INTERFACE & WEBPAGE

http://www.emptypaper.net/

http://www.aftermathreviews.com/

http://viljamis.com/

http://www.adhamdannaway.com/

http://www.adrianth.com/

http://watracz.com/

http://www.capitalawards.co.nz/

http://www.chichestermusicacademy.com/

http://www.alexnoren.com/

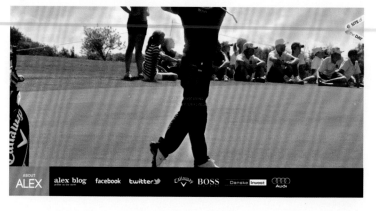

http://www.elegantseagulls.com/

http://www.eldodo.com.ar/

http://www.ericpaulsnowden.com/blog/

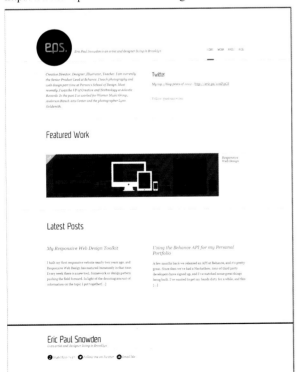

http://www.worldbank.org/

http://youdecide.bridgew.edu/

http://elevenmedia.com.au/

http://elementgraphics.co.uk/

http://www.usu.edu/

http://www.emilyridge.com/

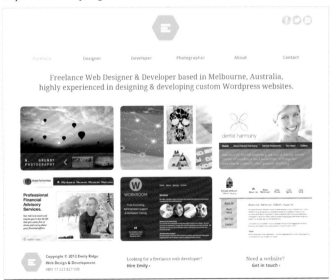

界面与网页设计 INTERFACE & WEBPAGE

电脑网页设计
企业网页

http://chutney.se/

http://costume.takami-bridal.com/index.html

http://hintongroup.com/

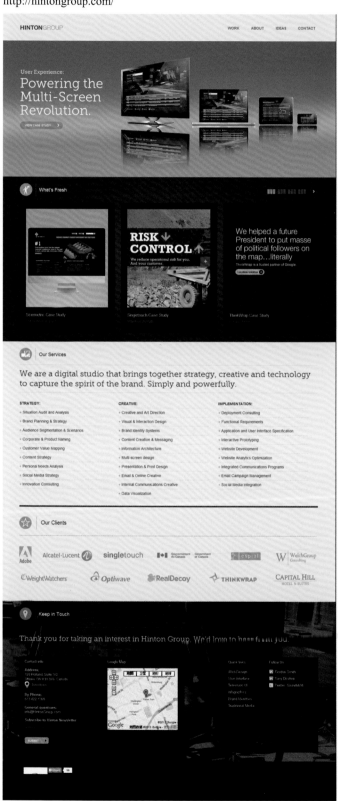

http://darford.com/

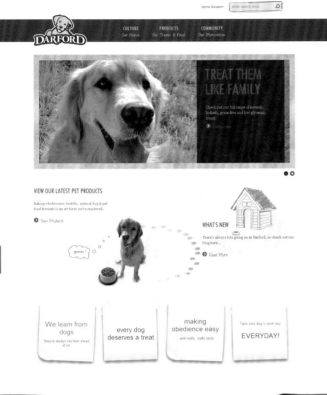

http://christmas.biltmore.com/#/day

http://www.barbaragallardo.com/

界面与网页设计 INTERFACE & WEBPAGE

http://feelandlive.com/

Con todos los sentidos

feelandlive es un pequeño estudio especializado en diseño y desarrollo web

Tres procesos ⌄

Análisis
Estudiamos tu proyecto. Realizamos un análisis inicial para ver las necesidades y empezamos con una arquitectura que defina las principales características.

Diseño
Creamos tu propia experiencia. Diseños a medida que transmiten más allá de las palabras y de los sentidos. Siempre siguiendo las últimas tendencias.

Desarrollo
Utilizamos las últimas tecnologías de desarrollo, siguiendo los estándares W3C. Nuestra especialidad es crear espacios funcionales y accesibles.

360°

Sentir / Feel
Sentimos internet, comunicamos 360 grados de sensaciones.

Vivir / Live
Vivimos internet, comunicamos utilizando ideas y experiencias.

Nos adaptamos ⌄

Nos adaptamos
iPhone / iPad / Android ...

El número de personas que accede a páginas web utilizando smartphones como el iPhone o tablets como el iPad, crece día a día.

Los tamaños de pantalla cambian según el medio con el que se accede. feelandlive realiza diseño web adaptable que se ajusta a todas las plataformas.

En Internet el tamaño sí importa.

Contáctanos ⌄

Cerca del mar ⌄

Cerca del mar
Costa Brava / Girona / Alrededores

Nos gusta hacer las cosas bien, ofrecemos toda nuestra ilusión y energía para realizar proyectos de calidad.

Nos movemos por la **Costa Brava, Girona** y **alrededores**, buscando proyectos en los que colaborar.

Si necesitas un producto de calidad pero a un precio razonable, ¿a qué esperas?

Contáctanos ⌄

Proyectos ⌄

Proyectos
Nuestros últimos trabajos

Tennis & Pàdel Piper's
El club de Tennis & Pàdel Piper's se fundó en 1966 y es uno de los primeros entornos donde se pudo practicar este deporte en la Costa Brava. Dispone de 6 pistas de tierra batida y 2 pistas de pàdel de máxima calidad.

Visitar la web ⌄

Tennis i Pàdel
a la Costa Brava

Agenda
El club al dia

Happy Life 55
El primer plan vacacional y residencial en el Caribe. Creado especialmente para mayores de 55 años, que hayan finalizado su vida laboral o dispongan de tiempo libre.

Visitar la web ⌄

+ 072

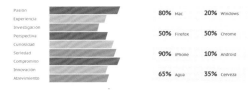

http://greenplanetsolutions.co.uk/

http://launchfactory.org/

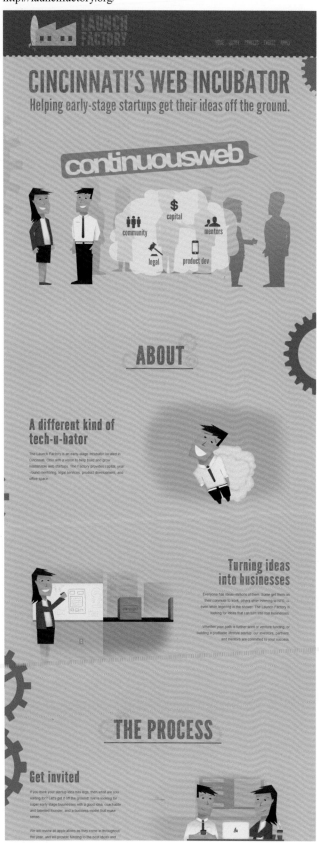

http://mtboucheriewinery.com/

http://madewithover.com/

HOW IT WORKS

Snap — Simply take a photo with your iPhone's camera

Type — Add custom typography to your photo with Over

Share — Let the world see your photo on Facebook, Twitter, Instagram, or Tumblr

MADE WITH OVER

WHAT PEOPLE ARE SAYING

"With Over, I'm thinking more like a publisher, with my photos telling stories instead of just being 'something to upload.' It's a great way to improve photos."
—Chris Brogan

"It is delightful, intuitive, and immediately easily useful. That's a winning combination for any app. I'm addicted - in a good way!"
—Aliza Sherman

"Simple. Intuitive. Innovative. And downright fun. I became a missionary for this app from the first moment I put words over my photos."
—Tim Sweetman

CONNECT WITH OVER

http://jt-roots.com/

http://jpn.nec.com/

http://www.digitalatelier.ro/#/work

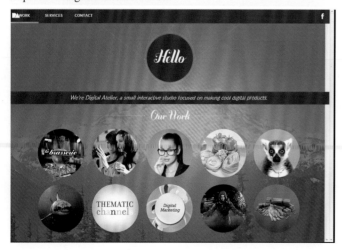

http://www.chrislinden.com/

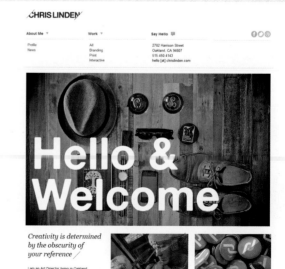

http://kai-kaga.jp/

http://morganguitars.com/

http://itinerary-app.com/

http://olivervogel.net/

http://realmacsoftware.com/

http://qui.lt/

http://rallyinteractive.com/

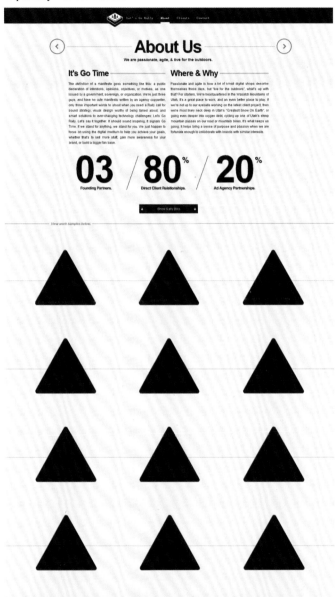

http://thegreetingfarm.com/

http://phobiahz.fr/

http://ot.nixon.com/

http://perconte.com/

界面与网页设计 INTERFACE & WEBPAGE

http://www.christhurman.com/

http://www.cornell.edu/

http://vivid-ness.co.uk/

http://www.alexlynnracing.com/

http://www.doubleglazingquote.net/

https://www.gov.uk/

http://www.comwerks.com/

http://www.brandify.co.uk/portfolio/

http://www.competa.com/

http://www.hayti.org/

http://www.bikingboss.com/

http://www.headlampcreative.com/

http://www.heinz.jp/

http://www.thismanslife.co.uk/

http://www.helmandafgn.com/

界面与网页设计 INTERFACE & WEBPAGE

http://www.factorycolors.net/

http://www.harvard.edu/

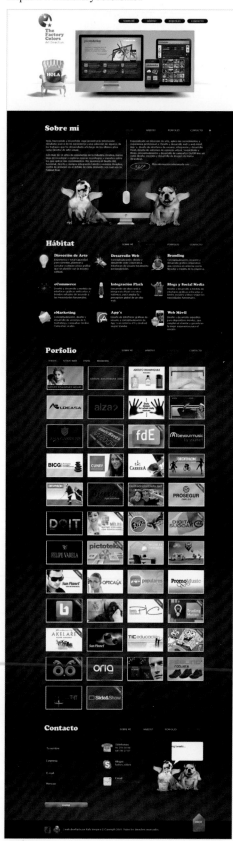

http://www.hiroogakuen.ed.jp/

http://www.hellomag.jp/

http://www.studiovs.nl/

http://www.thisistwhite.com/

http://www.sydneyspecialistdermatology.com.au/

http://www.storific.com/

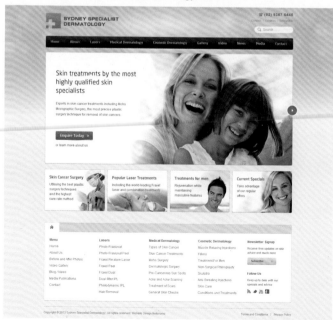

http://www.tinybigstudio.com/

http://www.sweatforacause.com/

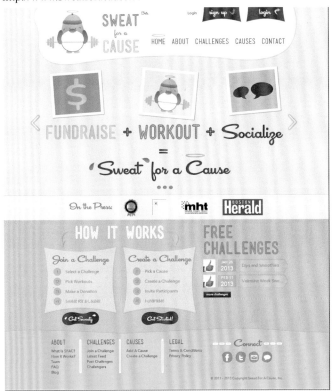

http://www.tku.ac.jp/

http://www.yyoga.ca/

http://www.vektorgrafik.se/

http://www.yessbmx.com/

http://www.vnsaga.com/

http://www.vaxx.cz/cs

http://b-porte.com/

http://www.yomeishu.co.jp/

http://www.honda.com.mt/

http://www.studiomoso.com.au/

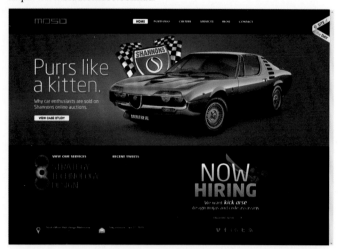

http://www.helveticbrands.ch/

http://www.suckyeah.com/

http://www.getaviate.com/

http://www.stinkdigital.com/en/

http://www1.umn.edu/twincities/index.html

http://www.style-story.com/

http://www.elegantbanners.com/

http://cookpad.com/

界面与网页设计 INTERFACE & WEBPAGE

电脑网页设计
电子商务

http://www.basketpascher.com/

http://www.kulthouse.com/

+ 092

http://labelsmag.com/

http://roaproduktion.se/

http://store.americanapparel.net/

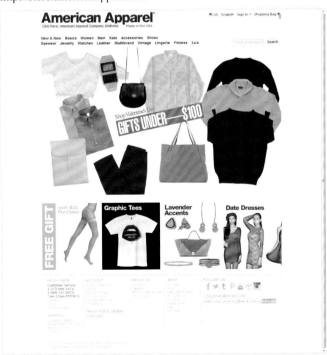

http://nibbuns.co.kr/

http://bagazimuri.com/

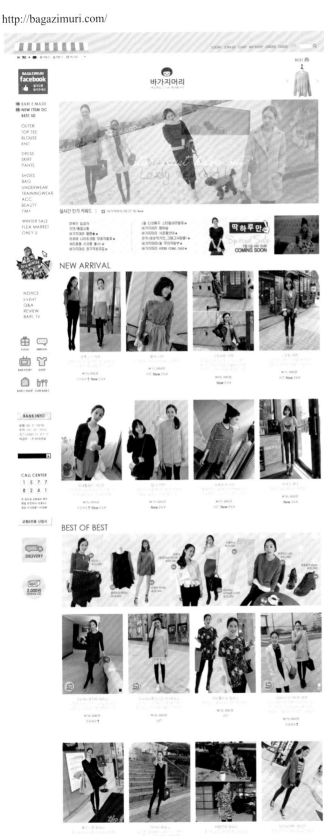

界面与网页设计 INTERFACE & WEBPAGE

http://www.qng.co.kr/

http://www.ddaddadda.co.kr/

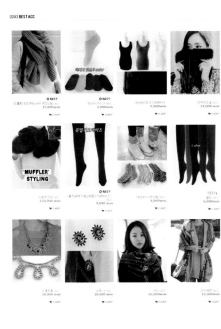

界面与网页设计 INTERFACE & WEBPAGE

http://imaginarydesign.co.uk/

http://tasteofink.com/

http://ohime.co.kr/

http://www.dclos.com/ca

http://www.uptowngirl.tv/

http://www.nationaltraveller.com/index.php

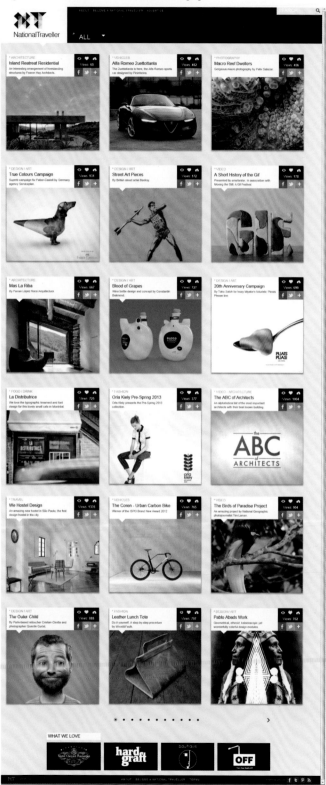

http://www.graceandtailor.co.uk/

http://www.waja.co.jp/

www.youaremusic.co.uk

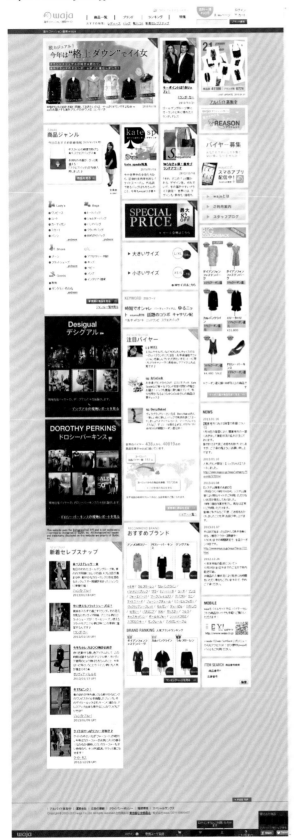

http://www.thaiscosta.com/

http://www.thebeardface.com/

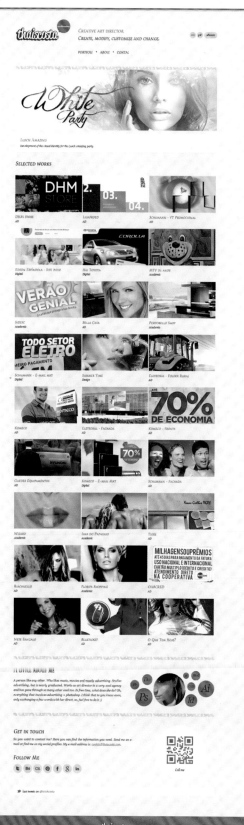
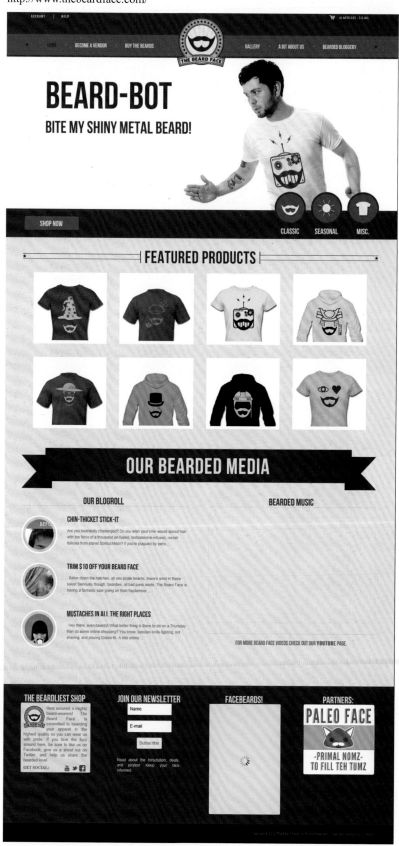

http://www.startupvitamins.com/

http://www.soiakyo.com/

http://www.ora-ito.com/

http://www.nenodesign.com/

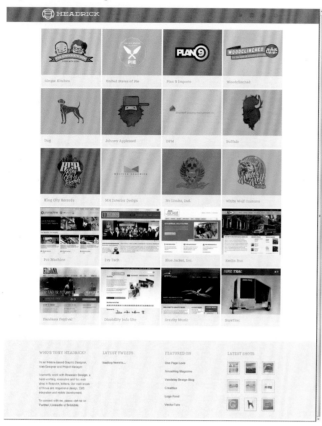

http://panoetic.com/

http://www.edsonespindola.com/

http://northernarmy.com/

http://www.coucoushop.ch/

http://www.noseinvaders.com/

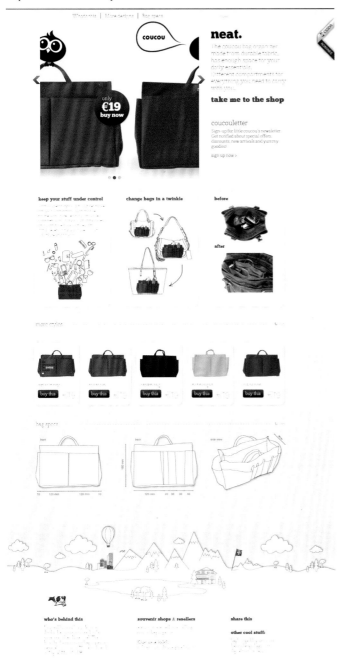

http://www.cleatskins.com/

http://www.siiimple.com/

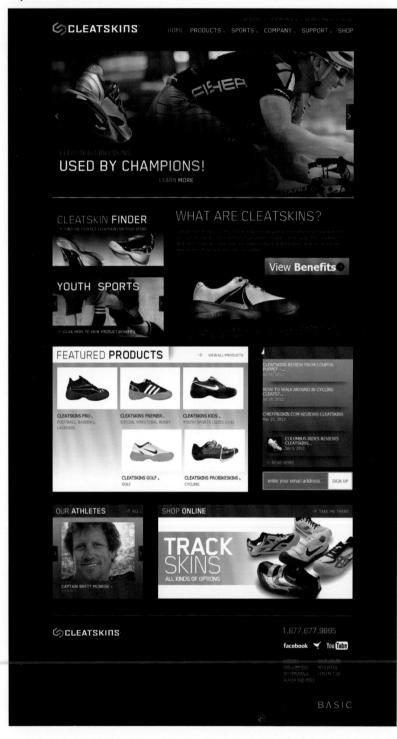

http://www.nhdist.com/

http://www.uzgames.com/

http://www.graceandtailor.co.uk/

http://www.extrememakovermeals.com/

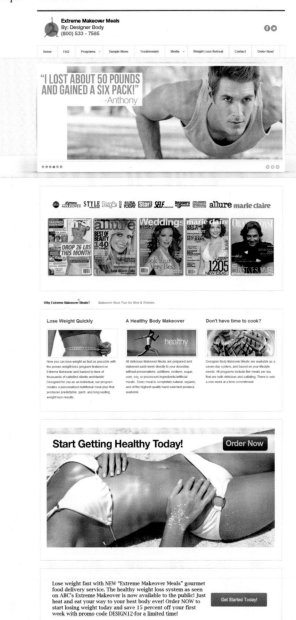

http://www.babosarang.co.kr/

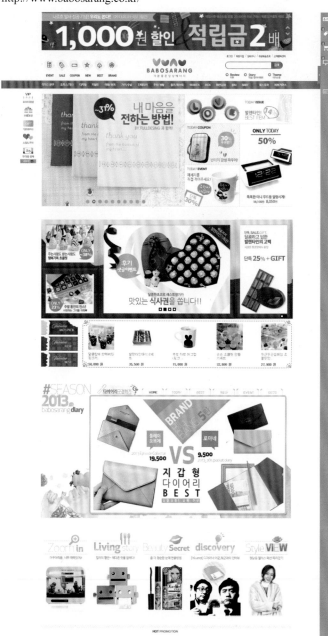

http://www.artmodic.com/

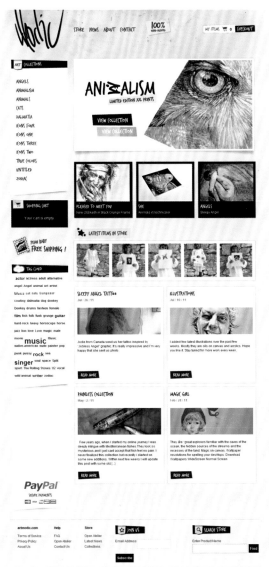

http://www.traveltomilano.com/

界面与网页设计 INTERFACE & WEBPAGE

http://www.babirolen.com/

http://thestylespy.com/

http://www.iamgaz.co.uk/

http://www.h3ostudio.com/

http://www.peterverkuilen.com/

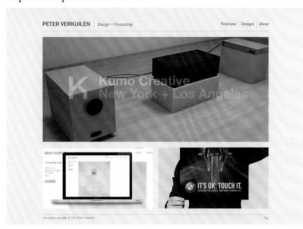

http://www.realmacsoftware.com/

http://www.thepurplebunny.com/

http://www.peterbristol.net/

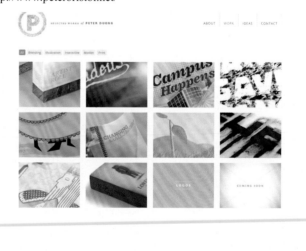

http://www.realviewapp.com/

http://www.resistenza.es/

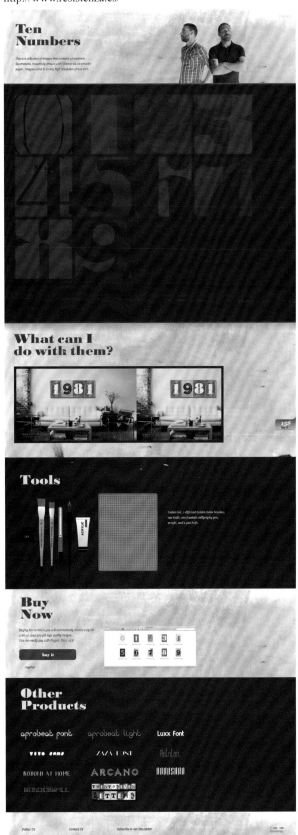

http://www.pedrolamin.com.br/2012/#inicial

http://www.resistenza.fi/

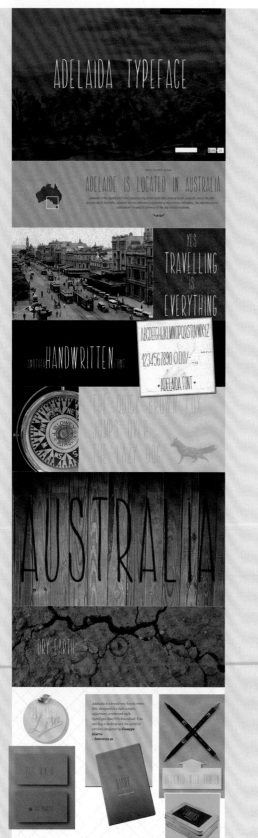
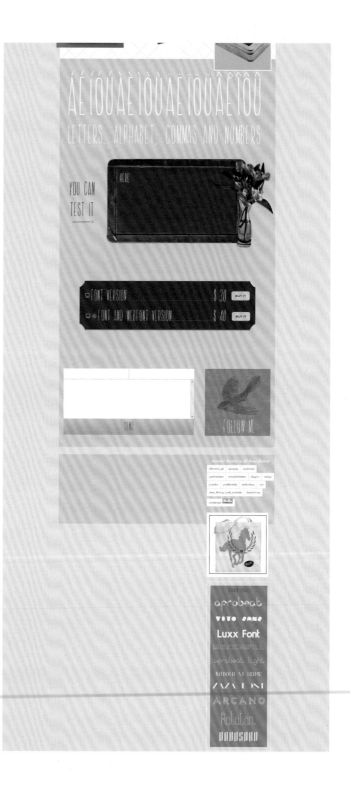

http://www.salveomamute.com.br/

http://www.seikansou.jp/

http://www.richardspierings.com/

http://www.remedycoffee.com/

http://www.shinwabank.co.jp/

http://www.sheenaoosten.com/

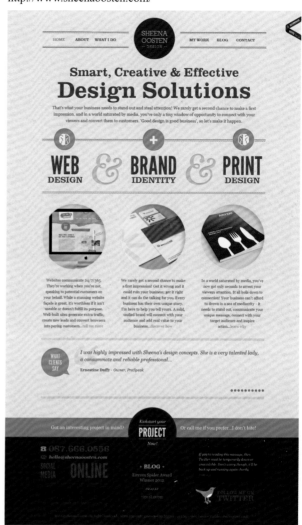

http://www.thedroidsonroids.com/

http://www.smurfett.com/

http://www.smithsonianchannel.com/sc/web/home

http://www.theflov.com/

http://www.snowbird.com/

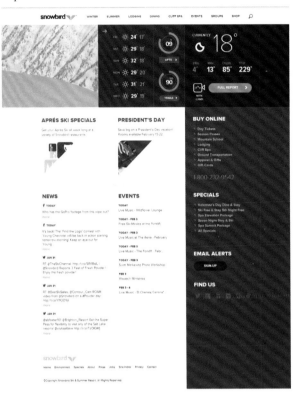

http://www.appuous.com/index.html

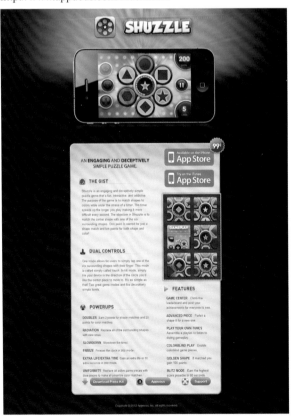

http://www.thewisdomofbees.com/

http://www.soulsurrender.com/

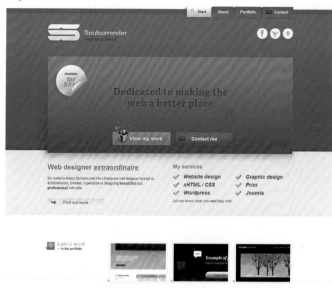

http://www.thevintagecateringcompany.com/

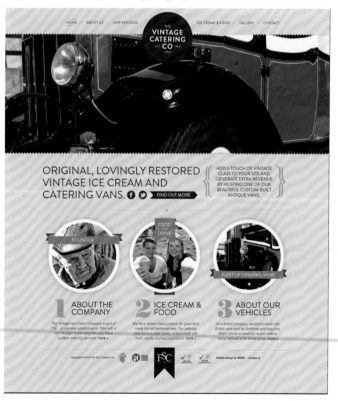

http://www.theseen.biz/

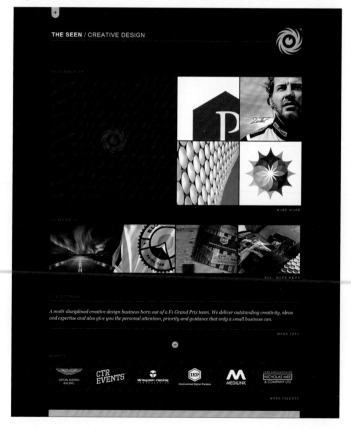

http://www.theblacksparrow.co.nz/

http://www.thecolorcure.com/

http://www.tdasset.co.jp/

界面与网页设计 INTERFACE & WEBPAGE

电脑网页设计
创意网页

http://dangerdom.com/

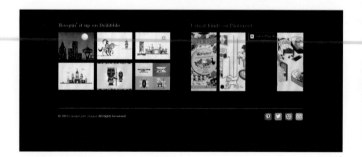

http://40digits.com/

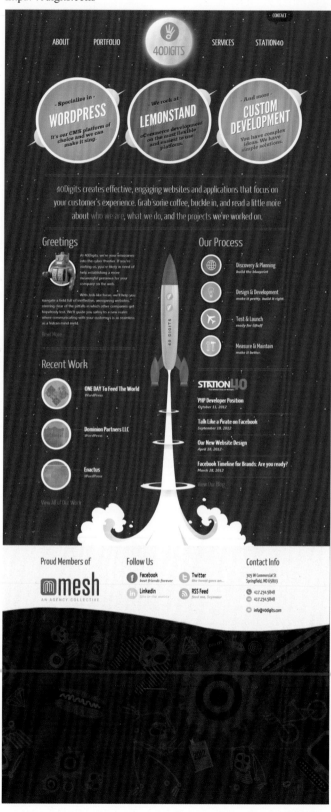

http://calobeedoodles.com/

http://cabedge.com/

http://chimpchomp.us/

http://www.creattica.it/

http://createdm.com/

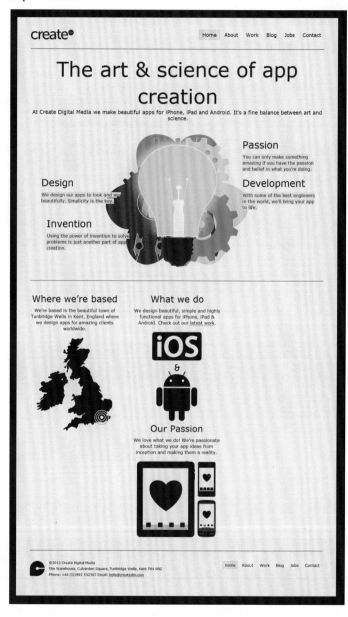

http://fairheadcreative.com/

http://www.clinttabone.com/

http://cobblehilldigital.com/

http://everytimeidie.net/

http://ds9creations.com/

http://www.cultstory.com/

http://hford.co.uk/

http://heathwaller.com/

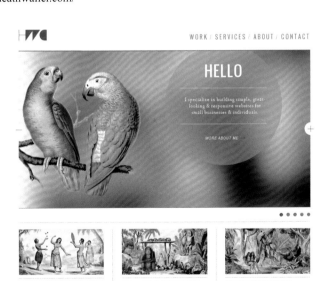

http://dezup.com/

http://hugs.fm/

http://inserviowebsolutions.co.uk/

http://kidoodleapps.com/

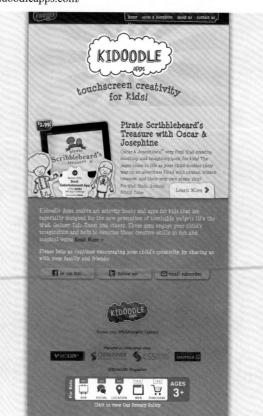

http://www.ivomynttinen.com/

http://hipinspire.com/

http://januarycreative.com/

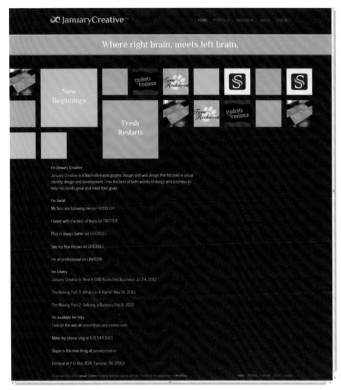

http://epicenterconsulting.com/

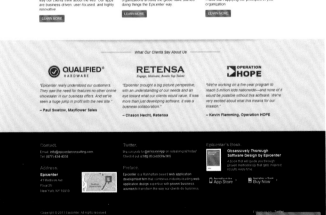

http://joshsender.com/

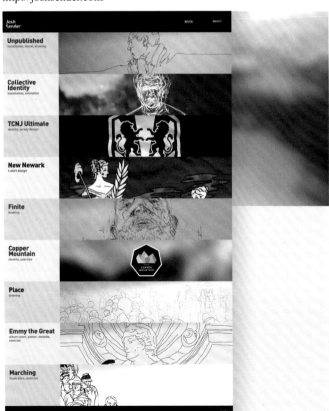

http://mizko.net/

http://leukocyt.com/

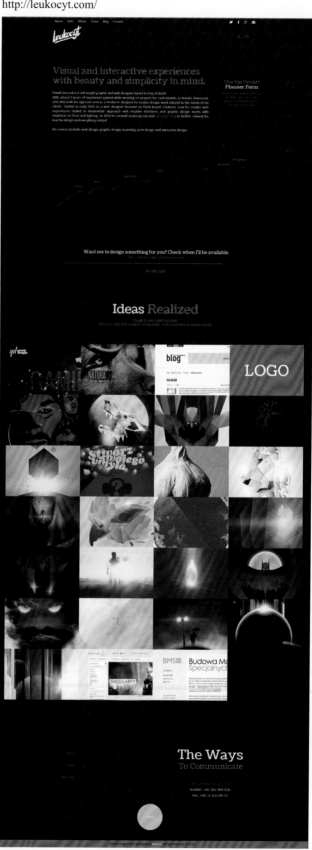

http://madebyshape.co.uk/

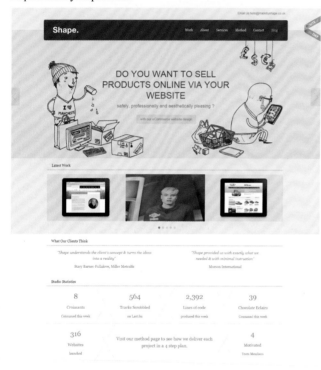

http://matowens.com/

http://www.marcovignolo.com/index.html

http://melonfree.com/

http://michaeldowell.com/

http://mikkoman.squarespace.com/personal/

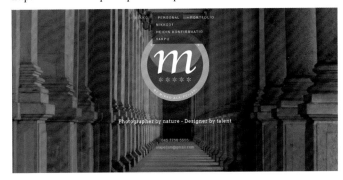

http://masswerks.com/

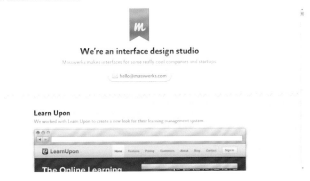

http://ricebowls.org/

http://schwartzandsonsny.com/

http://www.sloth.com/

http://simpleasmilk.com/

DESIGNERS OF WEB LOVERS OF TYPE KICKERS OF ASS

So you know we design for the web, love typography, and kick bottom, but what do we actually do? Our studio focuses mainly (but not solely) on designing for US based Tech-Startups. Most of our work focuses on building stunning, effective marketing sites, blogs and incredibly refined user experiences for web applications. We even develop them ourselves. We are in love with Ruby on Rails, WordPress and Illustration. We are a team of five extremely passionate and creative individuals who love what we do; and we would totally love to help you kick your competitors ass.

THE CREW

DAVID
David founded Simple as Milk in 2010 and is Co-Director. He is in love with typography, illustration and the startup culture. His design work brings brands to life, makes them vibrant, exciting and individual.

JAMES
James joined in 2011 and is now Co-Director here at SaM. He loves simplistic design with beautiful content layout. He loves increasing those conversion rates and also rocks out the majority of our Front-End Development.

KEVIN
Kevin is our developer. With over 7 years experience with Ruby, and 12 years with Web applications. Kevin takes our work and produces beautiful applications with amazing attention to detail.

GLEN
Glen is our creative mastermind. He comes up with stunning design concepts to help our clients stand out and to reinforce their brand. Like David he has a strong focus on illustration and typography.

SCOTT
Scott is our User Experience and Interaction Designer. He loves to create stunning interfaces and awesome interactions. He also revels in Front-End and jQuery development and, of course, loves typography.

JOBS?
No vacancies: We have never looked to hire, or had an open position, the right people for the job just seem to find us, fit perfectly within our family and stick around for the ride!

http://simpleasmilk.com/

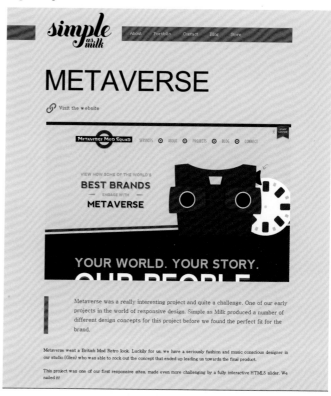

Metaverse was a really interesting project and quite a challenge. One of our early projects in the world of responsive design. Simple as Milk produced a number of different design concepts for this project before we found the perfect fit for the brand.

Metaverse went a British Mod Retro look. Luckily for us, we have a seriously fashion and music conscious designer in our studio (Glen) who was able to rock out the concept that ended up leading us towards the final product.

This project was one of our first responsive sites, made even more challenging by a fully interactive HTML5 slider. We nailed it!

http://snoop.ro/

http://www.cakesweetcake.co.uk/

http://www.bklynsoap.com/

http://thewikigame.com/

http://www.browserawarenessday.com/

http://storyboardsolutions.com/

http://www.bobvvs.se/

http://www.briansack.com/

http://www.blueacorn.com/

http://www.caffeine-creations.com/

http://www.carnationgroup.com/

http://www.colazionedamichy.it/

http://dubbedcreative.com/

http://www.csstardis.co.uk/

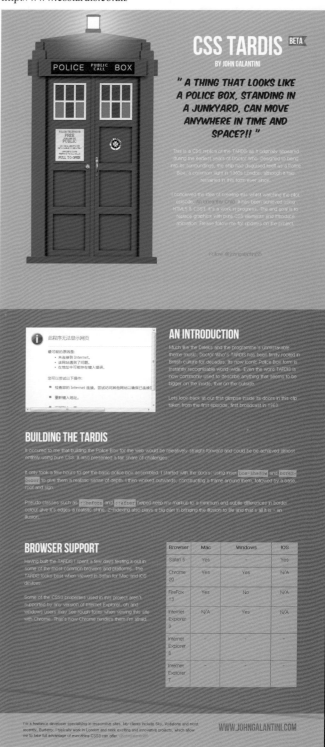

http://www.designzillas.com/

http://www.rockstarnewmedia.com/

http://creativestreamline.com/

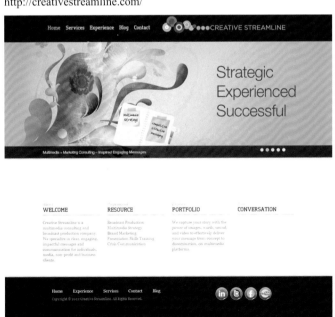

http://www.webbklubben.se/

http://copenhagen.chopeh.com/

http://www.eccentric-music.net/

http://www.i-am-tiago.com/

http://www.greenchameleondesign.com/

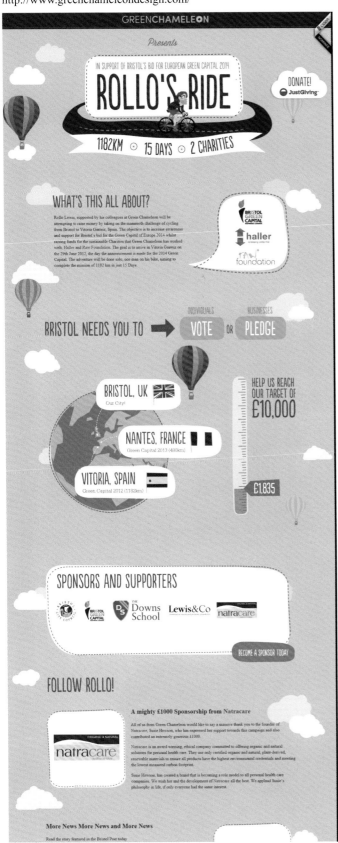

http://www.lifemusicfoundation.org/

http://www.krillbite.com/

http://www.logicdesign.co.uk/

http://www.idylliccreative.com.au/

http://www.ianjamescox.com/

http://www.krauppinc.com/

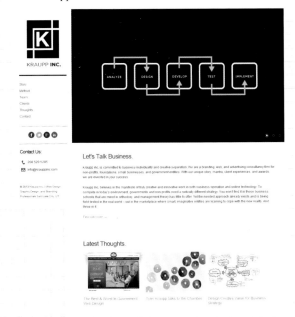

http://www.localwisdom.com/

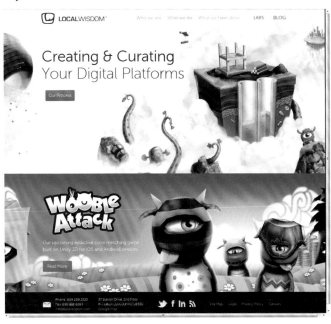

http://www.magpiart.com/

http://www.unicrow.com/?xk

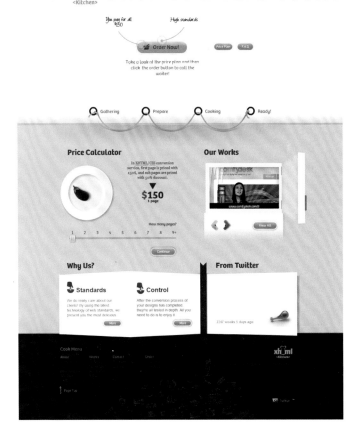

http://www.maisengasse.at/

http://www.madapple.cz/

http://www.musicatandem.com/

http://www.mitchdesigns.com/

http://www.monstersband.com/

http://www.mstrpln.com/landing.html

http://www.motownautomotiverepairs.com.au/

http://www.milliontrees.ca/

http://www.pixle.pl/

http://www.qubestudio.com/

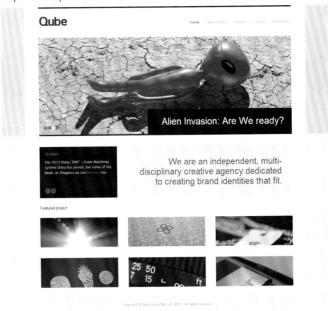

http://www.mywebsitebites.com/index/c/index/

http://www.traditionthreads.com/

http://www.tvlcorp.com/

http://www.tweetforlife.be/

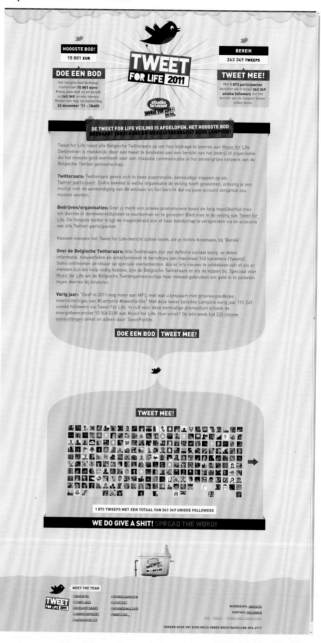

http://www.turningart.com/

http://www.muziekpark.nl/

http://www.vividcreative.com.au/

http://www.weroll.tv/

http://www.miramova.com/

http://www.plinestudios.com/

http://www.workdiary.de/

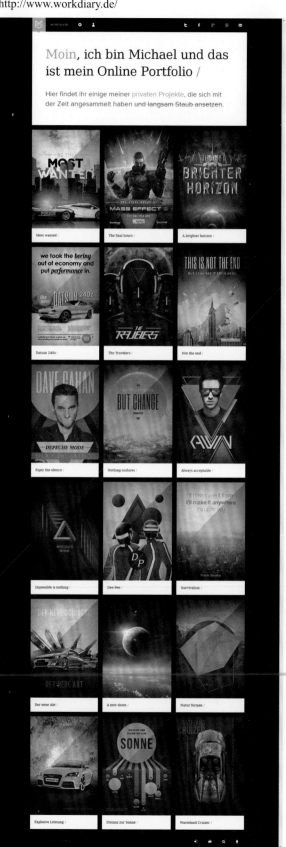

http://www.vivocha.com/

http://www.zync.in/

http://www.weblifting.at/

http://www.variousways.com/

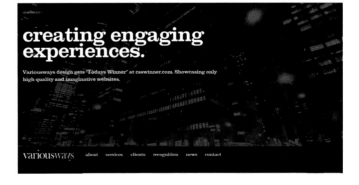

http://robedwards.org/

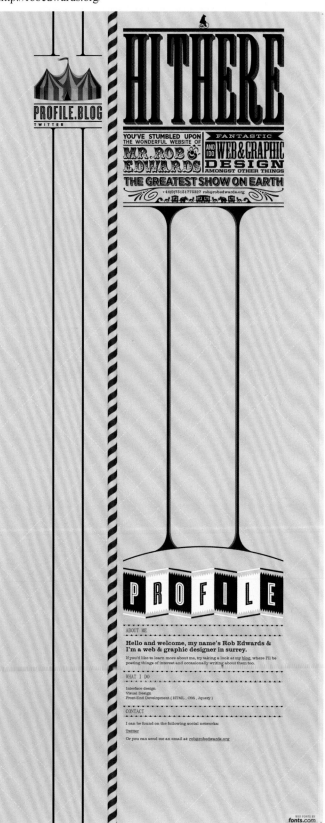

http://outofthisworld.co/

http://stillpointesanctuary.org/

http://show.james-oconnell.com/

http://tuispace.com/

http://tiradelhilo.es/

http://tobiaspersson.nu/

http://transmissioninc.com/

http://www.evoenergy.co.uk/

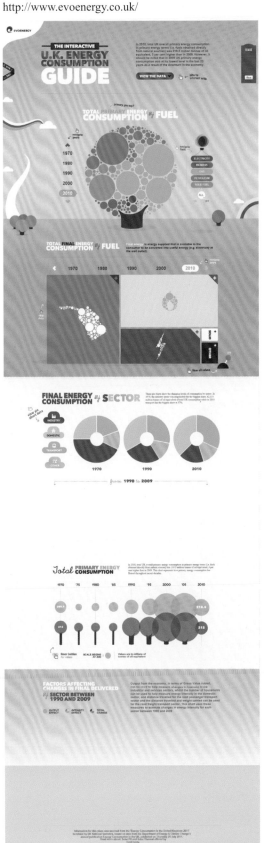

http://www.walletmap.com/

http://AAAAAAA.com/

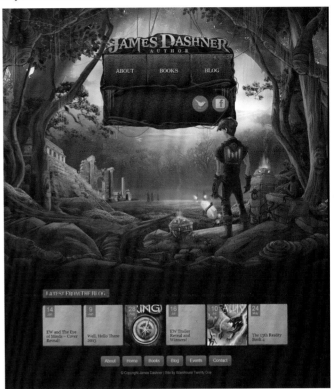

http://www.worryfreelabs.com/

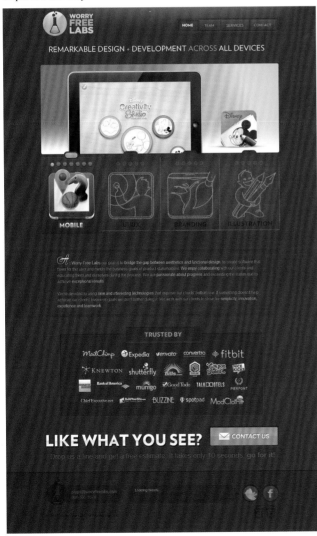

http://www.designergleb.com/

http://www.twofingerz.com/

界面与网页设计 INTERFACE & WEBPAGE

电脑网页设计
综合网页

http://www.hatbox.co/

http://listingscout.com/

http://davidbatra.se/

http://onemightyroar.com/

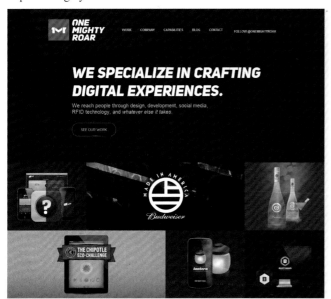

BRANDS

SHOW MORE

WORK

OUR PROCESS IS EDUCATED BY A BLEND OF YOUR GOALS AND OUR DISCOVERIES.

COMPANY

OUR COMPANY

PEOPLE

CONTACT
Any good relationship starts with a conversation, so go ahead... introduce yourself.

http://qlpros.com/

http://themeforest.net/

http://thelayoutlab.com/

http://thedarlingstarling.com/

http://www.evoenergy.co.uk/uk-energy-guide/

http://thehappybit.com/

http://www.3degreesagency.com/

http://wijmo.com/

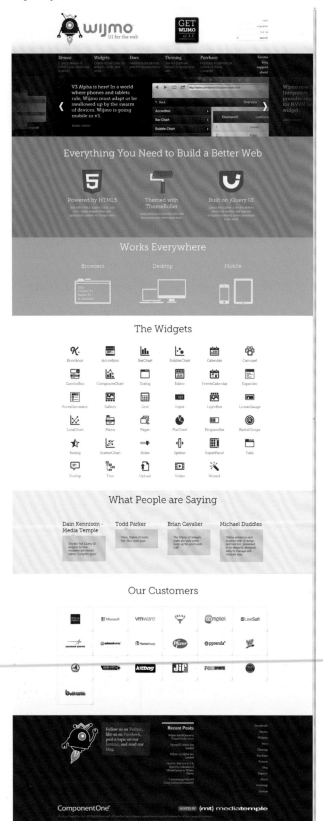

http://demo.krownthemes.com/

http://www.augustinteractive.com/

http://www.aaronvenn.com/

http://www.addictedtocoffee.de/

http://www.25fiftysix.com/

http://www.aaronrudd.co.uk/

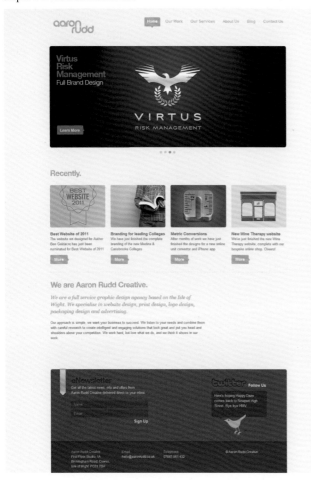

http://AAAAAAA.com/

http://www.cathedralcity.co.uk/

http://www.bememorable.co.uk/

http://www.tijuanaflats.com/

界面与网页设计 INTERFACE & WEBPAGE

http://www.captovate.com.au/

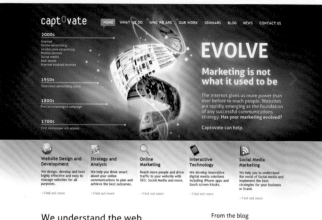

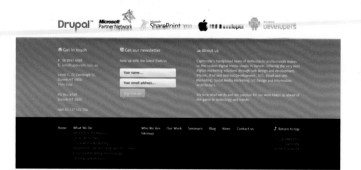

http://www.davinway.com/

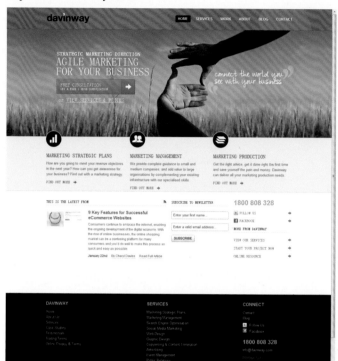

http://www.arizona.edu/

http://findjobswithstyle.com/

http://www.erikjankoopmans.nl/

http://www.fhoke.com/

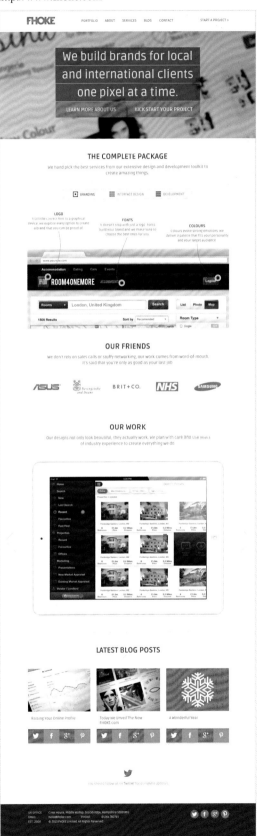

http://www.farmerjacks.co.uk/

http://www.jankoatwarpspeed.com/

http://www.decadentcakes.co.uk/

http://www.gazel.com/

http://www.kingsofmambo.com/

https://www.gathercontent.com/

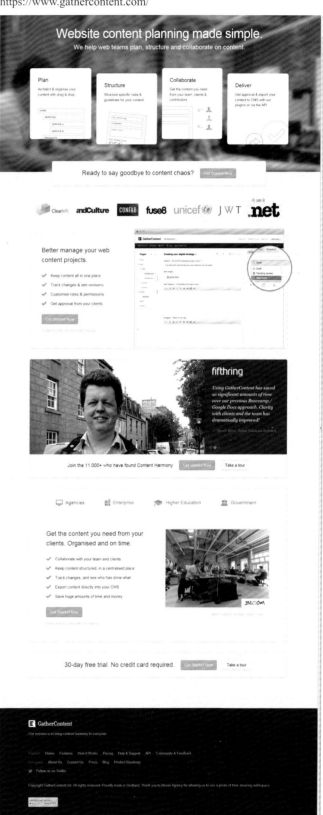

http://www.fcacampus101.com/

http://www.fluxar.com/

http://www.itsfirefly.com/

http://www.joescrabshack.com/

http://www.fhoke.com/

http://www.johnlikens.com/

http://www.johannes-sacht.de/

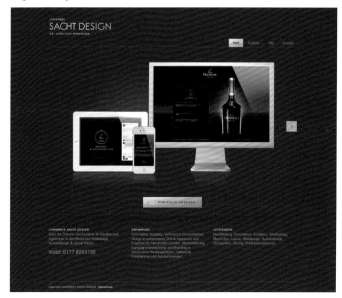

http://www.jochemgugelot.nl/

http://www.umpquaoats.com/

http://www.journeycreatives.com/

http://cheekymonkeymedia.ca/

http://AAAAAAA.com/

http://www.joselmerino.com/

http://www.kaleidoscope.com.au/

https://www.kanvess.com/

http://www.six11ink.com/

http://www.smart-j.com/

http://www.wallt.be/

http://www.viviledish.com/

http://www.decadentcakes.co.uk/

http://www.typographydeconstructed.com/

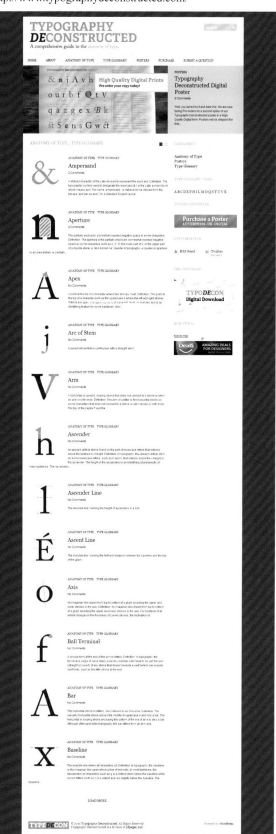

http://www.sitestitcher.com/

http://www.vivacitas.fr/

http://www.rocketlance.com/

http://www.roninapp.com/

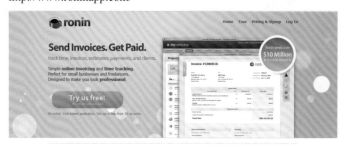